YOUR KNOWLEDGE HAS VALUE

Bibliographic information published by the German National Library:

The German National Library lists this publication in the National Bibliography; detailed bibliographic data are available on the Internet at http://dnb.dnb.de .

Imprint:

Copyright © 2015 GRIN Verlag, Open Publishing GmbH
Print and binding: Books on Demand GmbH, Norderstedt Germany
ISBN: 978-3-668-03776-2

This book at GRIN:

http://www.grin.com/en/e-book/305676/fluid-temperature-modelling-injected-at-surface-temperature-through-vertical

Jesus Rodriguez

Fluid temperature modelling injected at surface temperature through vertical wells

GRIN Publishing

GRIN - Your knowledge has value

Since its foundation in 1998, GRIN has specialized in publishing academic texts by students, college teachers and other academics as e-book and printed book. The website www.grin.com is an ideal platform for presenting term papers, final papers, scientific essays, dissertations and specialist books.

Visit us on the internet:

http://www.grin.com/

http://www.facebook.com/grincom

http://www.twitter.com/grin_com

FLUID TEMPERATURE MODELLING INJECTED AT SURFACE TEMPERATURE THROUGH VERTICAL AND DEVIATED WELLS. JESÚS ALBERTO RODRIGUEZ MORA[1].

ABSTRACT

This paper presents a brief bibliographic summary of the related well heat transfer models, most of them are designed for predicting heat loss along wells. All of those models lead to temperature profiles which show a lowering on fluid temperature during the injection from the wellhead to the wellbore, that make those models fit perfectly to the heat loss statement. The purpose of this article is to show a way to calculate the temperature increase of a fluid injected at surface temperature along a well by modifying an existing equation which was proposed by Boyun Guo (2004), this one satisfies the non-phase change of the injected fluid during its flow through the well, this implies that the selected model does not take into account latent heat change of the injected fluid, the previous fact is very important for this study because it is supposed that the injected fluid is not going to change its phase. Modifying Boyun Guo (2004) energy balance equation allowed to establish a profile temperature of a fluid injected at surface temperature into either a vertical or deviated well. It is also included a brief analysis of each of the parameters engaged on the process in order to determine the effect of those properties upon the injected fluid temperature.

Keywords: Fluid Temperature, Fluid at surface conditions, Modification, Temperature profile, Vertical and deviated wells.

1. Grupo de modelamiento de procesos hidrocarburos, GMPH, Escuela de Ingeniería de Petróleos, Universidad Industrial de Santander

RESUMEN

En el presente documento se muestra una revisión bibliográfica de los modelos relacionados con la transferencia de energía térmica en un pozo, muchos de ellos están diseñados para predecir la perdida de calor a lo largo del pozo. Todos estos modelos conducen a perfiles de temperatura que muestran una reducción en la temperatura del fluido durante la inyección del fluido desde cabeza de pozo hasta la cara del pozo, lo cual permite que estos modelos encajen perfectamente con el tema de perdida de calor. El propósito de este documento es mostrar una forma de calcular la temperatura de un fluido a lo largo del pozo, el cual es inyectado a temperatura ambiente modificando una ecuación existente, la cual fue propuesta por Boyun Guo (2004). La modificación de dicha ecuación permitió establecer un perfil de temperatura de un fluido inyectado a condiciones de superficie, esto aplica tanto para pozos verticales, como para pozos desviados. Adicionalmente se incluye un breve análisis de cada uno de los parámetros que intervienen en el perfil de temperatura con el fin de determinar el efecto de tales propiedades sobre la temperatura del fluido inyectado.

Palabras Clave: Temperatura de fluido, fluido a condiciones de superficie, Modificación, Perfil de temperatura, Pozos verticales.

WELL HEAT TRANSFER MODELS

The following is a compilation of models available, all of them associated to the well heat transfer phenomena.

Alves L.N. (1992), this model presented a forecast for the flowing temperature, this one is applied to surface pipelines, production and injection wells, one and two phase fluids, it is also applied to vertical and deviated wells. When developed, the flow was considered on steady state.

Belrute R.M. (1991), on this document was found the developing of a temperature profile simulator that shows the profiles during circulating and closing periods, considering a complex wellbore and various fluids inside it. The simulator was designed for mud and cement flow. It takes into account the existence of heterogeneous formations, affecting on different ways the temperatures of the flowing fluids.

Dawkrajai et al (2006), they proposed an specific condition that identifies water inputs considering the temperature profile on a horizontal well. In order to show the change on the fluid temperature they used a predictive model on different production conditions. On the developing of this model the difference between rock and fluid temperature was not considered.

Boyun Guo (2004), proposed three heat transfer correlations in order to predict the heat loss and the temperature profiles on insulated wells. To make the model as simply as possible the author took into account the highest resistance on the thermal system will suppress the others, considering just the insulating layer as the highest resistance. Despite the fact that there is no insulating layer for this study, the second highest resistance engaged on the phenomena is the fluid occupying the annular space, which will be used for modeling the temperature. One of the solutions is focused in steady state, two of them are for transient flow. Those equations could be carried out upon different conditions, considering a non-change phase fluid during the process.

Hagoort et al (2004), they analyzed the Ramey (1962) classic method, which was proposed for calculating the temperature on injection and production wells. They showed that this method has an approximation to the experimental results on long injection and production periods, shorter periods produce overestimated results on the temperature prediction.

Rajiv Sagar (1991), showed two methods to predict a two phase temperature profile on the production stage along the well. The first model is taken from an steady state energy equation, this one takes into account heat transfer mechanisms on the wellbore. The second model is a simplified version, which is based on the Coulter Bardon (1979) equation, whose equation includes the heat transfer mechanisms proposed by Ramey (1962) & Willhite (1967). It was not considered radiation and convection coefficients.

Ramey (1962), presented an approach to the heat transmission problem when hot and cold fluids are injected into the wellbore, this allows to calculate the fluid temperature inside tubing and casing based on time and well length. This model considers the heat resistances inside the wellbore between the circulating fluid and rock. This model as showed by Hagoort (2004) indicates that Ramey's model overestimate the temperature values on the earlier period of fluid injection.

Squier et al (1962), developed and solved a differential equation system which describes the temperature behavior when hot water is injected through a well. The model proposed describes the injection of hot water from surface to wellbore, the calculation is based on time and well length. This technique is applied to thermal recovery process, which means long injection periods. Compared to stimulation processes involving short periods of time, this model does not fit into the purpose of this study.

Tan X. et al (2011), developed mathematical models in order to simulate the temperature behavior along both vertical and horizontal wells, for controlling and evaluating acid stimulation on real time, this is done taking a temperature profile sequence on different periods of time, they consider an instantaneous temperature equilibrium between fluid and rock temperature.

Willhite (1967), selected specific methods to estimate the size of each component on the heat transfer process during hot water and vapor injection inside the well. The result was a methodology that allows the calculation of the global heat transfer coefficient (U). This value is considered as constant.

Yoshioka et al (2005), presented a model to predict the temperature profile on a horizontal well during steady state fluid production. The reservoir model is based on a correlation of mass and energy balance, considering a permeable and homogeneous enclosure. The model includes Joule Thomson effect, conductive and convective processes are also included. They also included the response when production rate, permeability and type of fluid are changed. They did not consider the formation fluid and the injected fluid velocity.

The analysis of the models above, led this study to take the one that fits properly into the objective of this research, that one is the model proposed by Boyun Guo (2004), this is the chosen one because it does not consider changes on the circulating fluid phase, it takes into account the well deviation angle, it is also easy to understand and manipulate. It considers just a resistance on the thermal energy transfer, this is the insulating layer covering the injection pipe.

WELL HEAT TRANSFER MODEL

On this section is shown the energy balance equation proposed by Boyun Guo (2004), the equations that represent the thermal energy loss on a fluid being injected, which are useful to calculate the temperature change on the hot fluid injected. Besides, it is also showed the modification to the original model, such change was applied on the overall energy equation modifying the equations that describe an increase on the temperature of the injected fluid.

MODEL FOR PREDICTING HEAT LOSS ON HOT INJECTED FLUIDS

The equation which represents the fluid heat loss will be explained using diagram 1 and the result is the equation 1.

Analyzing a section of the injected fluid (accumulation 2), it can be seen that this section losses thermal energy through the formation (heat flow "c") and also to the lower layer (accumulation 1), furthermore, the upper fluid layer (accumulation 3) gives thermal energy to the section analyzed. The previous analysis could be summarized on the next equation, which represents the heat accumulation on a section of the injected fluid at higher temperature than the surface temperature.

$$Accumulation\ 2 = q_b - q_a - q_c \qquad (1)$$

On the equation 2 it is shown the explicit solution for calculating the temperature of a hot fluid injected either a vertical or deviated well, that equation has been set to be on function of the constants α, β, γ, C, these constants are handled by equations 3, 4, 5 and 6, respectively.

$$T = \frac{1}{\alpha^2} \left[\beta - \alpha\beta L - \alpha\gamma - e^{-\alpha(L+C)} \right] \qquad (2)$$

Where

$$\alpha = \frac{2\pi R K_m}{v\rho_f CpsA} \qquad (3)$$

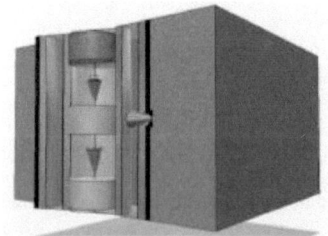

Diagram 1. Injection of a hot fluid.

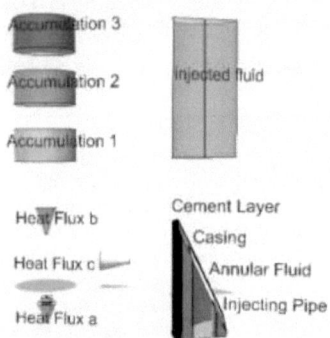

Diagram 2. Description of the parameters involved on the injection of a hot fluid.

$$\beta = \alpha GCOS(\theta) \tag{4}$$

$$\gamma = \alpha T_o \tag{5}$$

$$C = -\frac{\ln(\beta - \alpha^2 T_{fi} - \alpha\gamma)}{\alpha} \tag{6}$$

MODEL INTERPRETATION

The results produced by this model could be represented on a plot that correlates depth against temperature, as shown on the plot 1.

What is shown on the plot 1, is the injected fluid temperature loss, it is caused by the difference between the higher fluid temperature compared to rock temperature

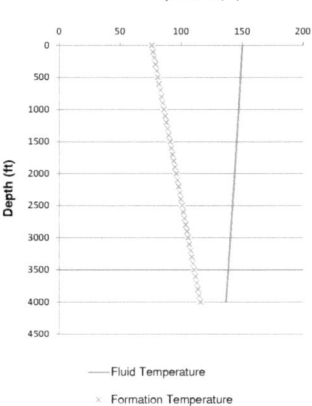

Plot 1. Formation temperature and injected fluid temperature.

MODEL MODIFICATION

On this section it will be show the changes made to the overall energy equation proposed by Boyun Guo (2004) which allows to find the explicit solution (equation 8) to determine the fluid temperature while it is injected at surface temperature, the new interpretation is summarized on the diagram 3 and 4.

The model has the next assumptions:

- Longitudinal heat transfer is ignored.
- Heat induced by friction does not have effect on the process.
- The pipe heat conduction capacity is much higher than the heat conduction capacity of the fluid inside the annular, which means that is the strongest resistance on the system on radial direction.
- The outer temperature is a lineal function of depth.

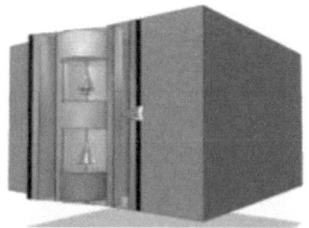

Diagram 3. Injection of a cold fluid

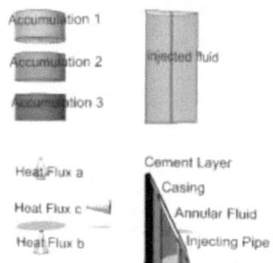

Diagram 4. Description of the parameters involved on the injection of a hot fluid.

Taking as reference diagram 3, both 'b' and 'c' heat flux give thermal energy to the 'accumulation 2', besides it can be assumed that the same accumulation gives thermal energy to 'accumulation 1', It is caused by the progress of the fluid inside the well which will get more and more thermal energy, the previous analysis is summarized on the equation 7.

$$Accumulation\ 2 = q_c + q_b - q_a \qquad (7)$$

It is shown the explicit solution for calculating the fluid temperature injected at surface temperature on the equation 8, which could be applied either vertical wells or deviated wells, this one depends on the following: α, β, γ and C, equations 9, 10, 11 and 12, respectively. All of them has changed as a result of the energy balance equation modification.

$$T = \frac{1}{\alpha^2}\left[-\beta - \alpha\beta L - \alpha\gamma + e^{\alpha(L+C)}\right] \qquad (8)$$

Where:

$$\alpha = -\frac{2\pi R K_m}{v\rho_f C p s A} \qquad (9)$$

$$\beta = -\alpha G COS(\theta) \qquad (10)$$

$$\gamma = -\alpha T_o \qquad (11)$$

$$C = \frac{\ln(\beta + \alpha^2 T_{fi} + \alpha\gamma)}{\alpha} \qquad (12)$$

MODEL RESPONSE

On this section will be shown the temperature profile resulting for both a water base fluid (Plot 2) and an oil base fluid (Plot 3), which are used on stimulation processes, the fluid properties that modifies the behavior of the fluid temperature are fluid heat capacity and fluid density.

The intake parameters to determine the water base fluid temperature profile are listed on the data sheet 1

Parameter	Value	Unit
Dtbg	7	Inches
Q	63,42	gpm
ρ_f	62,4	lb/ft^3
C_p	1	BTU/lb°F
s	1,875	In
km	0,611	BTU/hr°Fft
G	0,011	°F/ft
T_o	76	°F
Angle	0	Degrees
T_{fi}	76	°F
Casing inner diameter	10,75	In

Data sheet 1. Input parameters to calculate the temperature profile of a water base fluid.

It is observed on the plot 2 a temperature increase, which is the desired behavior on a fluid which is injected at lower temperature compared with the geothermal gradient, it is also seen that the fluid temperature remains below the rock temperature which gives a coherent response of the model.

It is observed the temperature behavior of an oil base fluid on the plot 3, the intake parameters are listed on the data sheet 2

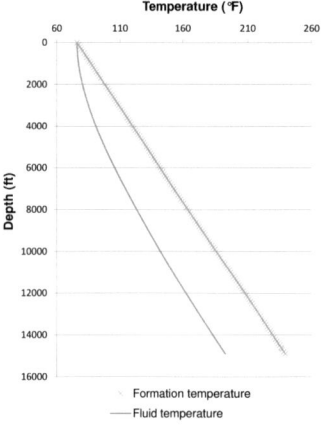

Plot 2. water base temperature profile, taking as input values those on the data sheet 1.

Parameter	Value	Units
Dtbg	7	Inches
Q	63,42	gpm
ρ_f	50	lb/ft^3
C_p	0,5	BTU/lb°F
s	1,875	in
km	0,611	BTU/hr°Fft
G	0,011	°F/ft
T_o	76	°F
Ángulo	0	grados
Ts	76	°F
Casing ID	10,75	in

Data sheet 2. Input parameters to calculate the temperature profile of an oil base fluid.

In order to make a comparison between the two of the fluids injected, the only intake parameters modified were: fluid density and heat capacity. Those properties are bounded

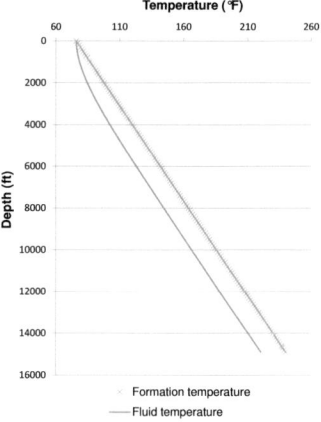

Plot 3. oil base temperature profile, taking as input values those on the data sheet 2.

directly to the fluid, both of them modify the temperature behavior, ignoring other fluid properties that do not have a strong effect on the temperature phenomena. What is observed on the plot 3 is a higher temperature on the oil base fluid compared with the water base fluid.

PARAMETER ANALYSIS ON THE MODEL

On this section is presented a brief analysis of the variables involved on the model, the intake parameters used on this analysis correspond to the ones on the data sheet 1, the purpose of this section is to show the impact of those values upon the temperature profile of a water base fluid.

The parameters which are going to be modified are: injection pipe diameter, flow rate, annular fluid thermal conductivity, well deviation angle and inner casing diameter, the results are shown on plots 4, 5, 6, 7 and 8.

Injection Pipe Diameter

Plot 4 shows an increase on the temperature of the injected fluid, it is caused by increasing the injection pipe diameter, this behavior is because the transversal area was magnified and it reduces the injected fluid velocity, which increases the residence time of the fluid on the pipe and so it generates a higher thermal energy transference.

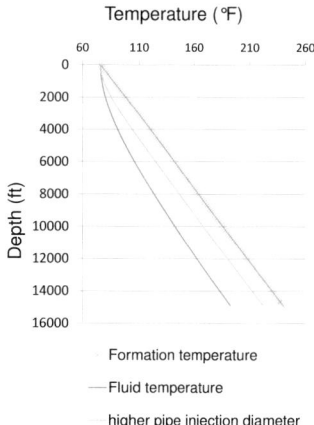

Plot 4. Higher Pipe Injection Diameter

Flow Rate

It is observed on plot 5 a decrease on the injected fluid temperature, it is caused by the reduction on the residence time of the fluid inside the injection pipe, and this is the opposite case of the injection pipe diameter. The previous behavior is caused by increasing the flow rate.

Annular Fluid Thermal Conductivity

The result of increasing the annular fluid thermal conductivity is a higher injected fluid temperature. It is because an increase on this property allows a faster heat transfer rate from

the rock to the fluid which is inside the injection pipe, it can be seen on plot 6.

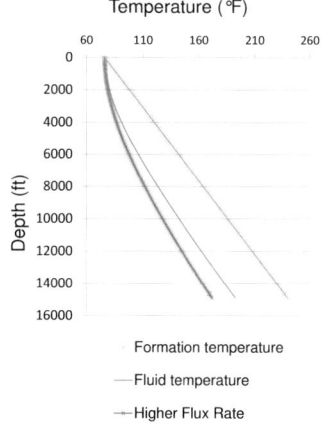

Plot 5. Higher Flux Rate

This property depends directly of the brine inside the annular space.

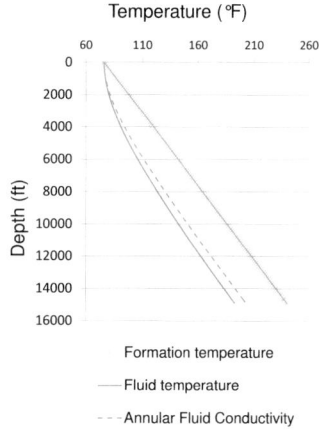

Plot 6. Higher Annular Fluid Conductivity

Well Deviation Angle

The increasing of the well deviation angle generates a reduction on the fluid injected temperature. This is caused by a higher inclination, which means a lower temperature per foot rate or geothermal gradient, this is show on the plot 7.

Inner Casing Diameter

When this property is increased the injected fluid temperature decreases, this is caused by a higher annular area where the brine is. This means an increase on the thermal resistance. The previous description can be seen on the figure 8.

A summary of the parameter analysis is shown on the data sheet 3.

models take into account changes on the fluid phase, which directly affects the latent heat of the injected fluid.

Parameter	Change on the original data	Temperature change
Injection pipe diameter	Increased	Increased
Flux Rate	Increased	Decreased
Conductivity	Increased	Increased
Deviation Angle	Increased	Decreased
Casing Diameter	Increased	Decreased

Data Sheet 3. Summary of the model behavior

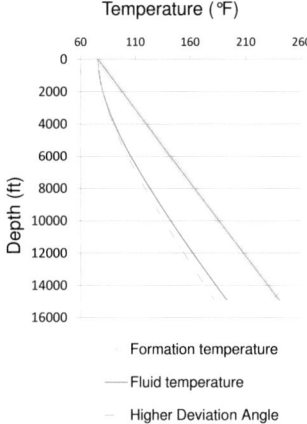

Plot 7. Higher Deviation Angle

CONCLUSIONS

The current models related to the thermal energy transfer inside a well are focused on the heat loss produced by the higher temperature of the injected fluid compared with the formation temperature, several of these

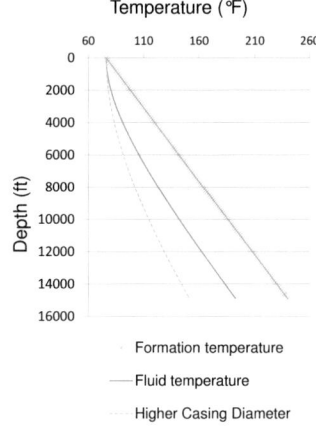

Plot 8. Higher Casing Diameter

The model that better fits on the purpose of this study is the one that considers a non-phase fluid change. Even though it is proposed for a hot fluid injection the model by Boyun Guo (2004) was the chosen one.

In order to adapt the model to the injection of a fluid injected at surface temperature a remake of the Boyun Guo's model (2004) was made by modifying the general energy equation, without removing the original parameters.

The result of modifying the balance energy equation, which is applied to fluids injected at surface temperature, gave as result coherent responses on the injected fluid temperature versus depth plots, even when the parameters involved on the phenomena where modified. One of the strongest reasons to confirm the coherence of the new version of the model was the fact that all the temperature profiles remain at lower temperature than the geothermal gradient, the second reason is that the injected fluid temperature tends to increase when it goes deeper inside the well.

Analyzing each of the parameters involved on the model produced as result the following: an increase on the injection pipe diameter and thermal conductivity of the fluid which is inside the annular space, affect directly the injected fluid temperature, making it to take higher values. On the other hand, when the next parameters are increased: flow rate of the injected fluid, well deviation angle and casing diameter, the injected fluid temperature will decrease.

NOMENCLATURE

q_a: Convective heat flux rate of the injected fluid.   (W)

q_b : Convective heat flux rate of the injected fluid.   (W)

q_c: Convective heat flux rate from the rock through the anular fluid.   (W)

T: Temperature.   (°C)

T_{fi}: Fluid temperature at wellhead.   (°C)

T_L: Fluid temperature at L distance.   (°C)

$T_{L+\Delta L}$: Fluid temperature at $L+\Delta L$ distance (°C)

T_o: Surface temperature.   (°C)

L: Longitudinal distance inside the well.   (m)

R: Outer radious of the injecting pipe.   (m)

K_m: Thermal conductivity of the anular fluid. (W/m-°C)

v: Fluid velocity inside the injection pipe. (m/s)

c_p: Specific heat of the injected fluid.   (J/Kg-°C)

s: Annular distance.   (m)

A: Transveral area of the injecting pipe.   (m²)

G: Geothermal gradient.   (°C/m)

$Dtbg$: Injecting pipe diameter.   (m)

Q: Volumetric rate flux.   (m³/s)

$Acumulation\ 1, 2\ y\ 3$:   energy   quantity accumulated inside each of the voume differentials (1, 2 y 3) per time unit.   (W)

θ: well deviation angle.   (Grades)

$\alpha, \beta, \gamma\ y\ C$: constants.   (non-dimmensional)

REFERENCES

1. ALVES, I.N.; ALHANATI, F.J.S. and SHOHAM, Ovadia. A Unified Model for Predicting Flowing Temperature Distribution in *Wellbores* and Pipelines. SPE Production Engineering, Volume 7, Number 4, November, 1992, p.363-367.

2. BELRUTE, R.M. A Circulating and Shut-in Well-Temperature-Profile Simulator. Journal of Petroleum Technology, Volume 43, Number 9, September, 1991, p. 1140-1146.

3. DAWKRAJAI, P, *et al*. Detection of Water or Gas Entries in Horizontal Wells From Temperature Profiles. SPE/DOE Symposium on Improved Oil Recovery, Tulsa, Oklahoma, USA, April, 2006. Paper.

4. GUO, Boyun; DUAN, Shengkai and GHALAMBOR, Ali. A Simple Model for Predicting Heat Loss and Temperature Profiles in Thermal Injection Lines and *Wellbores* With Insulations. SPE International Thermal Operations and

Heavy Oil Symposium and Western Regional Meeting, Bakersfield, California, March, 2004. Paper.

5. HAGOORT, Jacques. Ramey's *Wellbore Heat Transmission Revisited*. SPE Journal, Volume 9, Number 4, December, 2004, p.465-474.

6. RAMEY junior, H.J. *Wellbore Heat Transmission*. Journal of Petroleum Technology, Volume 14, Number 4, April, 1962, p.427-435.

7. SAGAR, Rajiv; DOTY, D.R. and SCHMIDT, Zelimar. Predicting Temperature Profiles in a Flowing Well. SPE Production Engineering, Volume 6, Number 4, November, 1991, p. 441-448.

8. SQUIER, D.P.; SMITH, D.D. and DOUGHERTY, E.L. Calculated Temperature Behavior of Hot-Water Injection Wells. Journal of Petroleum Technology, Volume 14, Number 4, April, 1962, p.436-440

9. TAN, X, *et al*. Measurement of Acid Placement with Temperature Profiles. SPE European Formation Damage Conference, Noordwijk, The Netherlands, June, 2011. Paper.

10. WILLHITE, G.P. Over-all Heat Transfer Coefficients in Steam and Hot Water Injection Wells. Journal of Petroleum Technology, Volume 19, Number 5, May, 1967, p.607-615.

11. YOSHIOKA, K, *et al*. A Comprehensive Model of Temperature Behavior in a Horizontal Well. SPE Annual Technical Conference and Exhibition, Dallas, Texas, October, 2005. Paper.

APPENDIX A.

On this section is shown the deduction of the equations that allow to calculate the temperature of the fluid injected at surface temperature.

$$q_a = \rho_f CpVAT_L \Delta t \tag{13}$$

$$q_c = \rho_f CpVAT_{L+\Delta L} \Delta t \tag{14}$$

$$q_b = 2\pi RK\Delta L \frac{dT}{dr} \Delta t \tag{15}$$

$$q_{cumulated} = \rho_f Cp\Delta L\Delta T \tag{16}$$

The horizontal temperature gradient is represented by:

$$\frac{dT}{dr} = \frac{(To+GLcos\theta)-T}{s} \tag{17}$$

The equation that represents the energy balance equation on the fluid differential inside the pipe is presented on the next equation:

$$q_{cumulated} = q_c + q_b - q_a \tag{18}$$

Including the equations (13), (14), (15) and (16) into equation (18), it is obtained:

$$\rho_f Cp\Delta L\Delta T = \rho_f CpVAT_{L+\Delta L}\Delta t + 2\pi RK\Delta L \frac{dT}{dr}\Delta t - \rho_f CpVAT_L\Delta t \tag{19}$$

Including $\rho CpVA\Delta t$ into equation (19), the next arrangement is get:

$$-\frac{dT}{dL} + \frac{2\pi RK}{\rho_f CpVA}\frac{dT}{dr} = \frac{1}{V}\frac{dT}{dt} \tag{20}$$

Including equation (17) into equation (20):

$$\frac{dT}{dL} + \frac{1}{V}\frac{dT}{dt} = \frac{2\pi RK}{\rho_f CpVA}\left[\frac{(To+GLcos\theta)-T}{s}\right] \tag{21}$$

The following parameters are get from equation (21):

$$a = \frac{-2\pi RK}{\rho_f CpVAs} \tag{22}$$

$$b = -aGCOS(\theta) \tag{23}$$

$$c = -aT_o \tag{24}$$

When flux is on steady state $\frac{dT}{dt} = 0$, replacing it into equation (21), the next is obtained:

$$\frac{dT}{dL} = \frac{aT}{V} + \frac{bX}{V} + \frac{c}{V} \tag{25}$$

From equation (25) the next constants are obtained:

$$\alpha = \frac{a}{v} \quad (26)$$

$$\beta = \frac{b}{v} \quad (27)$$

$$\gamma = \frac{c}{v} \quad (28)$$

Replacing and reorganizing (25) the following is get:

$$\frac{dT}{dX} = u = \alpha T + \beta L + \gamma \quad (29)$$

'u' makes easier the way of treating the model.

Taking 'T' as the dependent variable from equation (29):

$$T = \frac{u - \beta L - \gamma}{\alpha} \quad (30)$$

Deriving equation (30) the result is as follows:

$$\frac{dT}{dL} = \frac{1}{\alpha}\frac{du}{dL} - \frac{\beta}{\alpha} \quad (31)$$

From equation (29) it is obtained this $\frac{dT}{dL} = u$, and replacing it on equation (31) the following appears:

$$\frac{1}{\alpha}\frac{du}{dL} - \frac{\beta}{\alpha} - u = 0 \quad (32)$$

Reorganizing some of the items on the equation (32):

$$dL = \frac{du}{\beta + u\alpha} \quad (33)$$

Integrating (33) and including (29)

$$L + C = \frac{\ln(\beta + \alpha(\alpha T + \beta L + \gamma))}{\alpha} \quad (34)$$

In order to obtain the equation that represents the integration constant C, it is used next limit conditions on (34)

$$T = T_{fi} \ @ \ L = 0$$

$$C = \frac{\ln(\beta + \alpha^2 T_{fi} + \alpha\gamma)}{\alpha} \quad (35)$$

From equations (34) and (35) it is get the implicit equation that represents the fluid temperature gained when it flows vertically from wellhead to wellbore:

$$T = \frac{1}{\alpha^2}\left[-\beta - \alpha\beta L - \alpha\gamma + e^{\alpha(L+C)}\right] \quad (36)$$

YOUR KNOWLEDGE HAS VALUE

- We will publish your bachelor's and
  master's thesis, essays and papers

- Your own eBook and book -
  sold worldwide in all relevant shops

- Earn money with each sale

Upload your text at www.GRIN.com
and publish for free